NATIONAL GEOGRAPHIC KiDS

美国国家地理
双语阅读

Wolves

狼

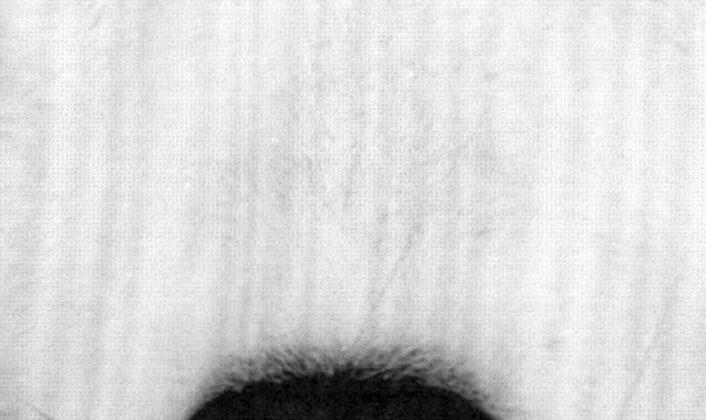

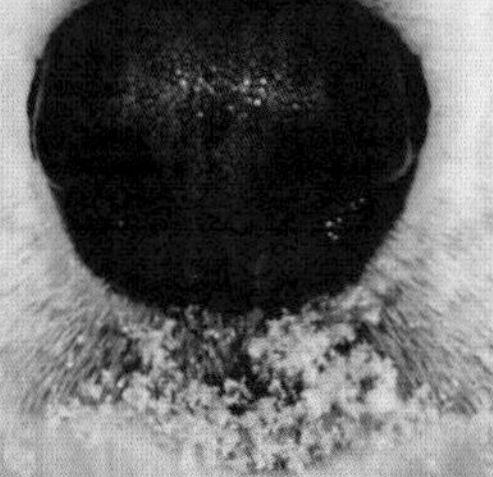

懿海文化 编著

马鸣 译

外语教学与研究出版社
FOREIGN LANGUAGE TEACHING AND RESEARCH PRESS
北京 BEIJING

京权图字：01-2021-5130

图书在版编目（CIP）数据

狼：英文、汉文 / 懿海文化编著；马鸣译. -- 北京：外语教学与研究出版社，2021.11（2023.8 重印）
（美国国家地理双语阅读. 第三级）
书名原文：Wolves
ISBN 978-7-5213-3147-9

Ⅰ. ①狼… Ⅱ. ①懿… ②马… Ⅲ. ①狼－少儿读物－英、汉 Ⅳ. ①Q959.838-49

中国版本图书馆 CIP 数据核字（2021）第 226244 号

出 版 人　王　芳
策划编辑　许海峰　刘秀玲　姚　璐
责任编辑　姚　璐
责任校对　华　蕾
装帧设计　许　岚
出版发行　外语教学与研究出版社
社　　址　北京市西三环北路 19 号（100089）
网　　址　https://www.fltrp.com
印　　刷　天津海顺印业包装有限公司
开　　本　650×980　1/16
印　　张　37.5
版　　次　2022 年 3 月第 1 版　2023 年 8 月第 4 次印刷
书　　号　ISBN 978-7-5213-3147-9
定　　价　188.00 元（全 15 册）

如有图书采购需求，图书内容或印刷装订等问题，侵权、盗版书籍等线索，请拨打以下电话或关注官方服务号：
客服电话：400 898 7008
官方服务号：微信搜索并关注公众号“外研社官方服务号”
外研社购书网址：https://fltrp.tmall.com

物料号：331470001

Table of Contents

What's That Sound?

Arrooooo!

There's a lonely howl in the distance. Then more voices join in. The chorus of howls sends a shiver down your spine.

What's making this spooky sound?

Wolves!

Wolves All Around

Wolves are found all over the world. They live in hot places like deserts. They also live in cold places like the North Pole.

The most common wolf is the gray wolf.

There are more than 30 kinds of gray wolves. And they are not just gray. They are brown, black, tan, and white, too.

Three Kinds of Gray Wolves

Iberian Wolf

Arctic Wolf

Timber Wolf

Wolves and Dogs

Wolves are the largest members of the dog family. Foxes, coyotes, jackals, wild dogs, and domestic dogs are also members of this family.

German Shepherd

Timber Wolf

Our pet dogs are relatives of the gray wolf. That's why they look alike.

Word Bite

DOMESTIC: Tame and kept by humans

This wolf and Golden Retriever are cousins!
longer snout
Wolf
stronger jaws
larger teeth
Dog
rounder head
shorter legs

Q What did the wolf say to his friend?

A Howl's it going?

But wolves and dogs are different in several ways.

Wolves have a longer snout, stronger jaws, and larger teeth. Dogs have a rounder head and shorter legs.

The biggest difference is that dogs like to be around people, and wolves would rather be around other wolves.

Wolves live in family groups called packs. A pack includes a male and female wolf, their young, and a few wolves that have joined from other packs.

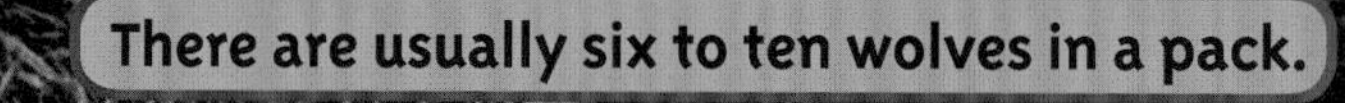
There are usually six to ten wolves in a pack.

Wolves need each other. Together they find food, protect one another, and care for their young. A wolf alone can't survive for long.

Hunting

Wolves are great hunters. They travel many miles without getting tired. They can usually run faster than their prey.

Wolves eat small animals such as rabbits. They also eat big animals such as moose, deer, caribou, elk, and bison.

And wolves eat a lot. They can each eat 20 pounds of meat in one meal. That's about 200 hot dogs!

Word Bite

PREY: An animal that is killed and eaten by another animal

Wolf Talk

How do wolves "talk" to the other wolves in their pack? They whimper, bark, growl, and snarl.

But when they need to talk long distance, wolves howl. And when one wolf starts howling, others tend to join in.

Howling is what wolves are famous for!

Watch out!
Help!
Where are you?
Yippee! We caught dinner!
Come hunting with me!
Stay off my turf!

Leaders of the Pack

The pack's leaders are called alpha wolves. There is one alpha male and one alpha female in each pack. They are the smartest and best hunters.

Alpha wolves guide the pack. They decide when to stop hunting and where to sleep at night. Alpha wolves also eat first at every meal.

Word Bite

ALPHA: A leader in a group

An alpha wolf lays its nose on top of a pack member's nose to show who's boss.

Pups

Baby wolves are called pups. Four to seven pups are usually born in each litter.

Pups weigh one pound at birth and can't see or hear. They snuggle safely in their den with their mother for the first two weeks.

Every day they grow bigger and stronger. At about three weeks old, the pups leave the den to explore.

Word Bites

LITTER: A group of animals born at one time

DEN: A hidden place in a cave or underground where animals live

When the pups are bigger, other wolves in the pack care for them, too. They bring the pups food. They also babysit them while the rest of the pack is hunting.

Q What do you call a lost wolf?

A A where-wolf!

Wolf pups start hunting with the pack when they are about four months old. When young wolves are two to three years old, they leave to form their own packs.

8 Wolf Wonders

1

Pups open their eyes when they are about two weeks old.

2

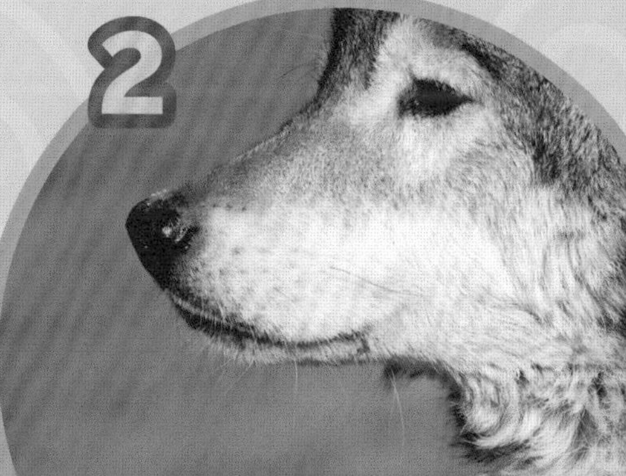

A wolf's sense of smell is about 100 times greater than a human's sense of smell.

3

Newborn pups can't keep themselves warm. They need mom for body heat.

4

Each wolf has its own howl, which sounds different from the howls of other wolves.

Q How do wolves eat?

A They wolf down their food!

5

Wolves usually won't hunt outside their own hunting grounds.

6

Pups play with "toys" such as a small dead animal from a kill, or a piece of its bone or fur.

7

An alpha wolf shows who's boss by walking tall with its tail and ears held high.

8

Wolves roam long distances—as far as 12 miles in one day!

Fewer Wolves

Wolves rarely attack people. They are afraid of them. But wolves do attack farm animals. Mostly for this reason, people have killed millions of wolves. So many wolves were killed that there were no longer any wolves in a lot of places.

Furs of wolves killed by hunters

Some people worried that wolves might become extinct. So they decided to help.

They passed laws to protect wolves. Today wolves are returning to many places around the world.

Word Bite

EXTINCT: A type of plant or animal no longer living

Wolves Return

Wolves have returned to Yellowstone National Park. Once the park had many wolves. But only one wolf was left by 1926.

Why did the wolf want the chicken to join the band?

He wanted drumsticks!

Scientists brought wolves from Canada into Yellowstone in 1995. The wolves had pups.

Now there are about 100 wolves in Yellowstone. Once again wolves make their home in the park.

Stump Your Parents

Can your parents answer these questions about wolves? You might know more than they do!

Answers are at the bottom of page 31.

1

Where do wolves live?

A. In the desert
B. In the mountains
C. In the forest
D. All of the above

2

Wolves like to "talk" long distance to one another by ____________.

A. Chirping
B. Squealing
C. Howling
D. Buzzing

3

What is the most common kind of wolf?

A. Red wolf
B. Gray wolf
C. Italian wolf
D. Werewolf

4

What is the leader of a wolf group called?

A. Head wolf
B. President
C. Alpha wolf
D. Chief

5

What is a baby wolf called?

A. A cub
B. A calf
C. A pup
D. A foal

6

What is the main reason there are fewer wolves today?

A. Hunters
B. Pollution
C. Hurricanes
D. Disease

7

What do wolves like to eat?

A. Hamburgers and french fries
B. Berries and grasses
C. Insects
D. Rabbits, deer, and elk

Answers: 1) D, 2) C, 3) B, 4) C, 5) C, 6) A, 7) D

Glossary

ALPHA: A leader in a group

DEN: A hidden place in a cave or underground where animals live

DOMESTIC: Tame and kept by humans

EXTINCT: A type of plant or animal no longer living

LITTER: A group of animals born at one time

PREY: An animal that is killed and eaten by another animal

▶ 第 4—5 页

那是什么声音?

嗷!

远处传来一声孤独的嗥叫。接着更多声音加入进来。这样的齐声嗥叫让你脊背发凉。

这令人毛骨悚然的声音是什么发出来的?

狼!

▶ 第 6—7 页

世界各地的狼

世界各地都有狼。它们生活在炎热的地方，比如沙漠。它们也生活在寒冷的地方，比如北极。

最常见的狼是灰狼。

世界上有三十多种灰狼。而且它们不仅仅有灰色的，也有棕色的、黑色的、褐色的和白色的。

三种灰狼

伊比利亚狼

北极狼

森林狼

▶ 第 8—9 页

狼与狗

狼是犬科动物家族中数量最多的成员。狐狸、郊狼、豺、野犬和家犬也是这个家族的成员。

我们的宠物犬是灰狼的亲戚。所以它们长得很像。

第 10—11 页

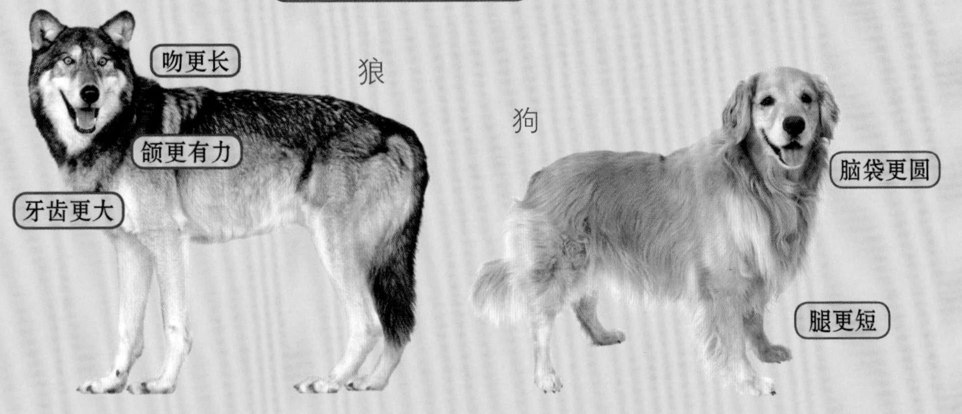

但是狼和狗在好几个方面都不一样。
狼的吻更长，颌更有力，牙齿更大。狗的脑袋更圆，腿更短。
最大的不同是狗喜欢和人待在一起，而狼更喜欢和别的狼待在一起。

第 12—13 页

狼群生活

狼过着群居生活，组成“狼群”。一个狼群里有一只公狼、一只母狼和它们的幼崽，还有从别的狼群加入进来的狼。

一个狼群里一般有6到10只狼。

狼需要彼此。它们一起觅食，相互保护，照料幼崽。独狼不能生存很久。

第 14—15 页

捕猎

狼是捕猎高手。它们即使走很远也不会累。它们通常比猎物跑得快。

狼吃小动物，比如兔子。它们也吃大动物，比如驼鹿、鹿、北美驯鹿、美洲赤鹿和北美野牛。

而且狼的食量很大。它们一顿可以吃掉 20 磅（约 9.07 千克）肉。那相当于大约 200 个热狗！

狼群小词典

猎物：被另一只动物杀死并吃掉的动物

▶ 第 16—17 页

狼的对话

狼如何与狼群里别的狼“对话”呢？他们呜咽，吠叫，低吼，咆哮。

但是，远距离对话时，狼会嗥叫。当一只狼开始嗥叫时，别的狼也会加入进来。

嗥叫是狼的成名技！

▶ 第 18—19 页

狼群的首领

狼群的首领被称为“头狼”。每个狼群都有一个雄性头领和一个雌性头领。它们是最聪明、最优秀的猎手。

头狼指挥狼群。它们决定什么时候停止捕猎，晚上睡在哪里。每顿饭也都是头狼先吃。

头狼把鼻子放在狼群其他成员的鼻子上面，来表明谁是“老大”。

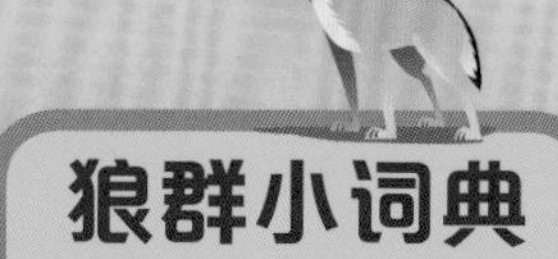

狼群小词典

头领：一个群体的首领

▶ 第 20—21 页

狼崽

狼宝宝被叫作“狼崽”。狼通常一窝生 4 到 7 只狼崽。

狼崽刚出生时只有 1 磅（约 0.45 千克）重，看不见东西，也听不到声音。出生后的前两周，它们和妈妈一起待在安全的兽穴里。

每天它们都会长得大一些、壮一些。差不多三周大时，狼崽就能爬出兽穴探险了。

狼群小词典

一窝： 同时出生的一群动物

兽穴： 隐蔽在山洞里或地下、动物生活的地方

▶ 第 22—23 页

狼崽长大一些之后，狼群里别的狼也会照顾它们。它们会为狼崽带回食物。当狼群里的其他成员外出捕猎时，它们也会像妈妈那样照料狼崽。

大约四个月大时，狼崽开始和狼群一起捕猎。幼狼长到两三岁时，它们会离开，组建自己的狼群。

第 24—25 页

关于狼的 8 件奇事

1

大约两周大时，狼崽才睁开眼睛。

2

狼的嗅觉比人类的嗅觉灵敏大约100倍。

3

刚出生的狼崽无法维持自身的体温。它们需要妈妈帮忙维持体温。

4

每只狼都有独属于自己的嗥叫声，听起来与别的狼的嗥叫声不一样。

5

狼通常不会到自己的捕猎区之外捕猎。

6

狼崽玩“玩具”，比如被杀死的小动物，或者它的骨头或毛皮。

7

头狼走路时把尾巴和耳朵竖得高高的，以表明谁是“老大”。

8

狼每天都走很远——可达12英里（约19.31千米）！

狼越来越少

狼很少袭击人类。它们害怕人类。但狼会袭击农场上的动物。主要是出于这个原因，人类已经杀死了上百万只狼。那么多的狼被杀死，以至于很多地方都没有狼了。

被猎人杀死的狼的毛皮

如今，野外生活着大约40只赤狼。

有些人担心狼会灭绝，因此他们决定帮助狼。他们通过了法律来保护狼。现在，狼又回到了世界上的许多地方。

狼群小词典

灭绝：一种植物或动物不再存在

第 28—29 页

狼回来了

狼已经回到了黄石国家公园。这个公园里曾经有很多狼。但是到 1926 年，只剩下一只狼了。

1995 年，科学家把狼从加拿大带到了黄石国家公园。狼生了狼崽。现在黄石国家公园里有大约 100 只狼。狼再一次在公园里安了家。

▶ 第 30—31 页

挑战爸爸妈妈

你的爸爸妈妈能答出这些有关狼的问题吗？你知道的可能比他们还多呢！答案在第 31 页下方。

1

狼在哪里生活？

A. 在沙漠里　　B. 在大山里

C. 在森林里　　D. 以上都是

2

狼喜欢通过 ______ 与别的狼远距离“对话”。

A. 啁啾声　　B. 鸣叫声

C. 嗥叫声　　D. 嗡嗡声

3

最常见的狼是哪一种？

A. 赤狼　　B. 灰狼

C. 意大利狼　　D. 狼人

4

狼群的首领叫什么？

A. 首狼　　B. 总统

C. 头狼　　D. 酋长

5

狼宝宝叫什么？

A. 幼兽　　B. 牛犊

C. 狼崽　　D. 马驹

6

现在狼越来越少的主要原因是什么？

A. 猎人　　B. 污染物

C. 飓风　　D. 疾病

7

狼喜欢吃什么？

A. 汉堡包和炸薯条　　B. 浆果和草

C. 昆虫　　D. 兔子、鹿和美洲赤鹿

第 32 页

词汇表

头领：一个群体的首领

兽穴：隐蔽在山洞里或地下、动物生活的地方

家养的：驯化的，并被人类饲养

灭绝：一种植物或动物不再存在

一窝：同时出生的一群动物

猎物：被另一只动物杀死并吃掉的动物